AF894678

International Trends in Patents, Research Funding and Technology for Unmanned Aerial Systems

A summary of recent developments

by: John S. Palmisano Reach John at: palmisano@gmail.com

ISBN 978-1-57440-646-7 © 2020 Primary Research Group Inc.

Table of Contents

Table of Contents 2

Introduction: Briefing on Unmanned Aerial Systems 4

Recent History 4

Major Categories of UAS Use 5

Overarching Development Drives and Funding Pushes for UAS Technology 6

Focus of UAS Research 6

Counter-Drone Research 7

UAV Fuel Cells 7

Licensing and Patents 8

Trends in Patents, by Country 9

Major Chinese Patent Holders for UAS 10

Major US Patent Holders in UAS 10

European Patent Holders in UAS 12

Government Grants and Research Institutions 14

Public Funding of Non-Military UAS in the United States 14

UAS Startups & University-Private Company Joint Ventures in UAV 25

Military Funding and Research 28

OVERALL DOD FUNDING FOR UAV'S 28

Air Force Funding Plans 31

Marine Corp Funding 34

US Operations Special Command Funding 35

DOD Counter-drone Budget 35

EU Government Spending 36

Chinese Drone Research and Spending 38

University Research on UVA in China 38

Private Industry Research of UVA in China 39

Public Sector Research on UVA Technology in China 41

China's Role in UAV Scientific Publishing 42

China's Military Drones 43

South Korean Drone Research and Spending 46

Korean Research on UAS 46

Korean Government Ministries Responsible for Drone Development 47
Public Agency Drone Research 48
South Korean Army Research on Drones 49
Japanese Drone Research and Spending 51
Japanese University and Public/Private Research 51
Japanese Private Industry Drone Research 52
Japanese Military Spending on Drones 53
About the Author 54
Bibliography 55

Introduction: Briefing on Unmanned Aerial Systems

Recent History

Historically, development into practical unmanned aerial systems (UAS)[1] began soon after manned flight. Significant research funding however did not pour in until the first successful 'smart bombs' of WWII, such as Nazi Germany's V-1 autonomous "buzz bomb" and the Henschel Hs 293 "Fritz X" anti-ship radio controlled glide bomb.

While UAS have been in constant development ever since, the very recent renaissance in UAS began in the 2000's with two primary triggers. The first is the major successes of the MQ-1 Predator drone and it's later more heavily armed MQ-9 Reaper cousin, leading to a rapid increase in military funding for UAS. The second major growth trigger was a rapid blossoming of the small hobbyist community specializing in robotics and micro-drones. This growth, combined with rapidly dropping costs for UAS-related electronics, led to an explosion of interest and startups producing all modern consumer grade drones in use today.

---

[1] There is no unified name for unmanned aerial systems (UAS). Other common terms are drone, unmanned aerial vehicle (UAV), and quadrotor.

Major Categories of UAS Use

The major uses of UAS technology can be roughly divided into three major categories: consumer, industrial, and military. The consumer category uses small and cheap drones for inspection, video, scientific, humanitarian, and hobbyist purposes. The industrial category uses larger heavy lift or higher altitude drones for mapping, package transport, and agriculture. The military category is self-explanatory.

There are nearly endless uses for drones currently in use: package delivery, medicine delivery, crop dusting, imaging, inspection, tracking wildfires, counting wildlife, search and rescue, tracking sharks near beaches, crop yield estimation, mapping, archeology, airport bird control, water testing, film and video, window cleaning, and many more. Other imagined future uses include public transportation, aerial construction, ambulance drones, and internet access distribution.

Overarching Development Drives and Funding Pushes for UAS Technology

Focus of UAS Research

The majority of federal funding distributed for developing new drone technologies is earmarked for military systems, followed by scientific research, and finally economic tech-development programs. Individual states and localities are using drone funding programs to boost local tech communities and to diversify away from dependence on older traditional industries, such as oil and gas.

Current UAS technology development drives are focused on:

- Power sources: batteries, fuel cells, solar power, and draft-catching gliders
- Control systems: autonomy, collaboration, and swarm behaviors
- Sensing: imaging, obstacle avoidance, and target identification
- Communication techniques: ad-hoc networking
- Data collection processing: of imagery and video
- Counter-drone technology: perimeter security and military

Additional active research into how drones interact with the public are:

- Drone safety
- Recharging methods

- Package delivery methods
- Legal frameworks
- Counter-drone technologies

Counter-Drone Research

As concerns grow around the potential security threats drones may pose to both civilian and military entities, a new market for counter-drone technology is rapidly emerging. Counter-drone technology (counter-UAS or C-UAS), refers to systems that are used to detect and/or neutralize an UAS. At least 235 counter-drone products have been identified, produced by 155 manufacturers in 33 countries. The most popular drone detection techniques are radar, RF detection, EO, and IR. The most popular interdiction technique is jamming. [1]

UAV Fuel Cells

Major organizations working on fuel cell-powered UAVs include the US Naval Research Laboratory, NASA, South Korea's Hylium Industries Inc., Turkey's Marmara Research Centre Energy Institute and National Boron Research Institute, Cella Energy and the University College London in collaboration with the Scottish Association for Marine Science and Arcola Energy, INHA University (South Korea), Lockheed Martin, and Silent Falcon UAS Technologies. A survey review of fuel cells for drones may be had here: [2].

Licensing and Patents

In 2019 the Danish Technological Institute released a comprehensive report on world-wide UAS patents. Most information and charts below were derived from this report. You can download it here: [3].

The report identified a total of 28,398 unique UAS patent families and 3757 companies with UAS-related patents, with the overwhelming majority of these patents filed after 2012 (fig 2.3 below). More than 82% of all UAS patents have been published after 2016.

FIGURE 2.3 PUBLICATION TRENDS, 1990-2018

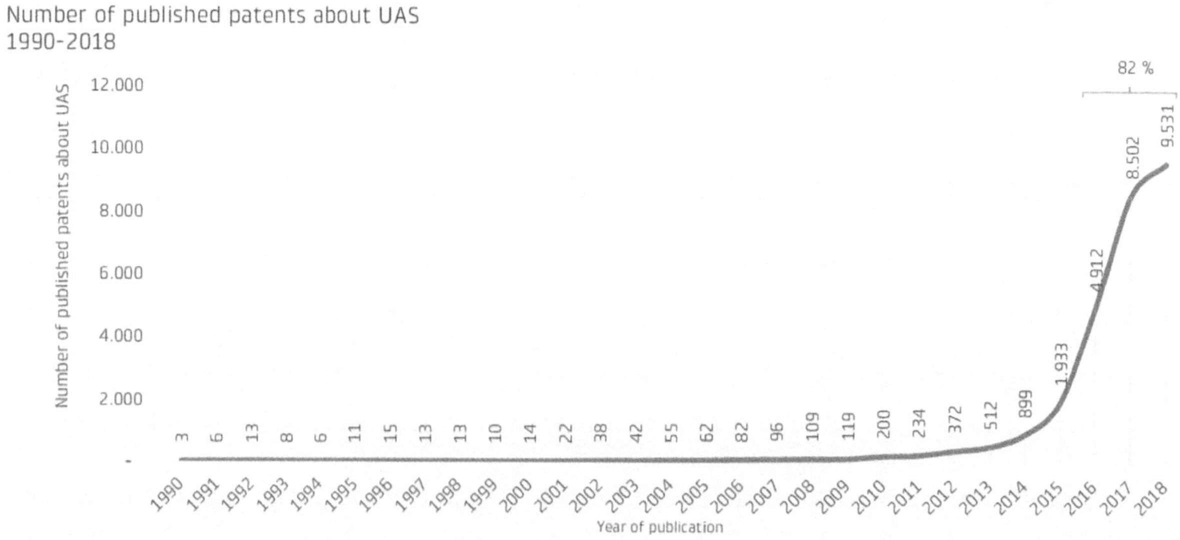

Source: Calculations by Danish Technological Institute based on a search in global patent databases via PatSnap. **28.398 patent families** identified. Syntax: TAC:("Unman*" $w1 "Aircraft*") OR TAC:("remote*" $W1 "aircraft*") OR TAC:("Micro Air Vehicle*") or TAC:("Unman*" $w1 "Aerial*"). Extract 7. dec. 2018.

Trends in Patents, by Country

The US has been issuing UAS patent publications since 1992 and dominated the field until 2008 when it was surpassed by China. Since then, China is by far the most UAS patented country in the world. The US is a distant second, and all other remaining countries combined represent only a small sliver of total published UAS patents (figure 2.4 below). US patents are concentrated at the east and west coasts as well as around Chicago, Arkansas, and Texas. Both in the US and China, the UAS industry is traditionally based on military and industrial companies, but new actors such as Amazon and Walmart are also visible. South Korea also has a notable number of patent assignees. Beyond Chinese, American, and South Korean patenting companies, very few companies in other countries have taken out UAS patents.

FIGURE 2.4 PUBLICATION TREND BY COUNTRY, 1990-2018 (MOST PATENT BY TOP 10 COUNTRIES)

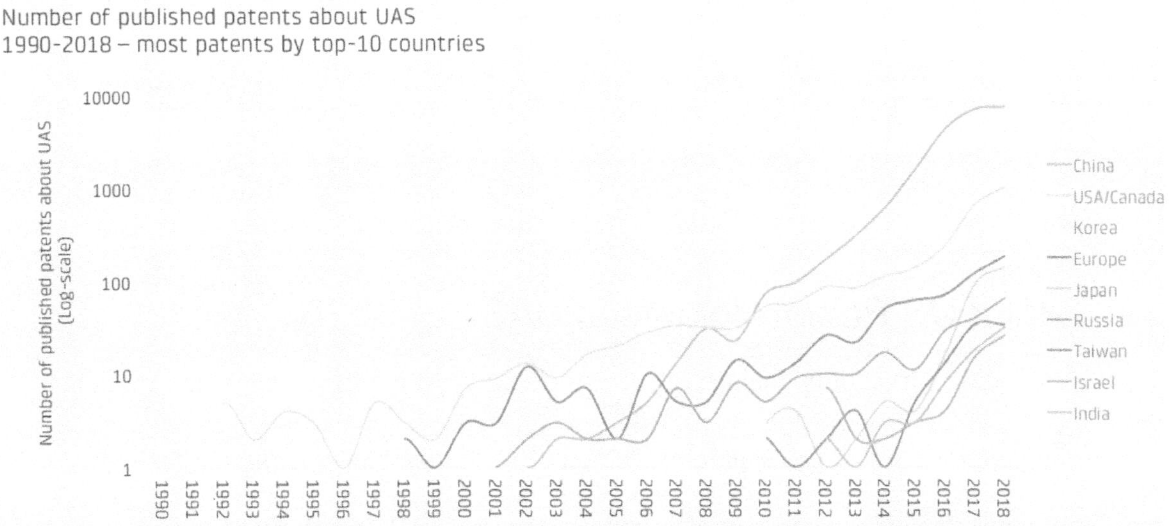

Source: Calculations by Danish Technological Institute based on a search in global patent databases via PatSnap. **28.398 patent families** identified. Syntax: TAC:("Unman*" $W1 "Aircraft*") OR TAC:("remote*" $W1 "aircraft*") OR TAC:("Micro Air Vehicle*") or TAC:("Unman*" $w1 "Aerial*"). Extract 7. dec. 2018. Figures for Europe calculated separately.

Major Chinese Patent Holders for UAS

The main location for UAS patents is currently China, with Shanghai and Hong Kong leading the number of patent assignees. Shenzhen Daijang Innovation Tech (DJI, https://www.dji.com/), a leading UAS manufacturer for aerial videography and industrial high grade drones, has filed the most patents and has around 800 patents related to UAS. See figure 2.6, below.

FIGURE 2.6 TOP 10 GLOBAL COMPANIES WITH UAS-RELATED PATENTS

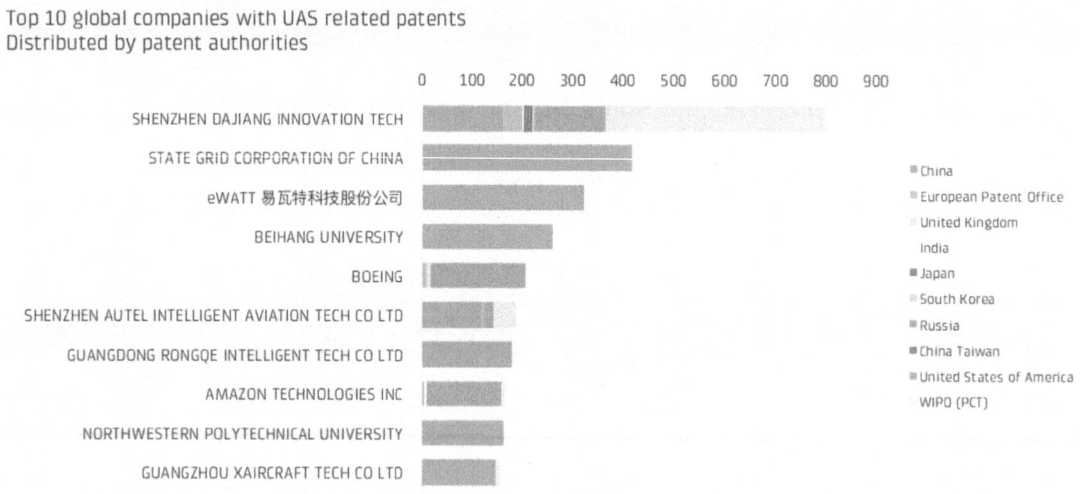

Source: Calculations by Danish Technological Institute based on a search in global patent databases via PatSnap. **2.873 patent families** identified. Syntax: TAC:("Unman*" $w1 "Aircraft*") OR TAC:("remote*" $W1 "aircraft*") OR TAC:("Micro Air Vehicle*") or TAC:("Unman*" $w1 "Aerial*"). Extract 7. dec. 2018. Figures for top 10 companies calculated separately.

Major US Patent Holders in UAS

Excluding China-only patents, the second-ranking patent filing company, far behind China's DJI, is Boeing, which is closely followed by Amazon Technologies Inc., and Honeywell as the fourth-ranking company (see figure 2.10 below).

FIGURE 2.10 NUMBER OF PATENTS ONLY FROM TOP 10 PLAYERS (EXCLUDING CHINA-ONLY PATENTS)

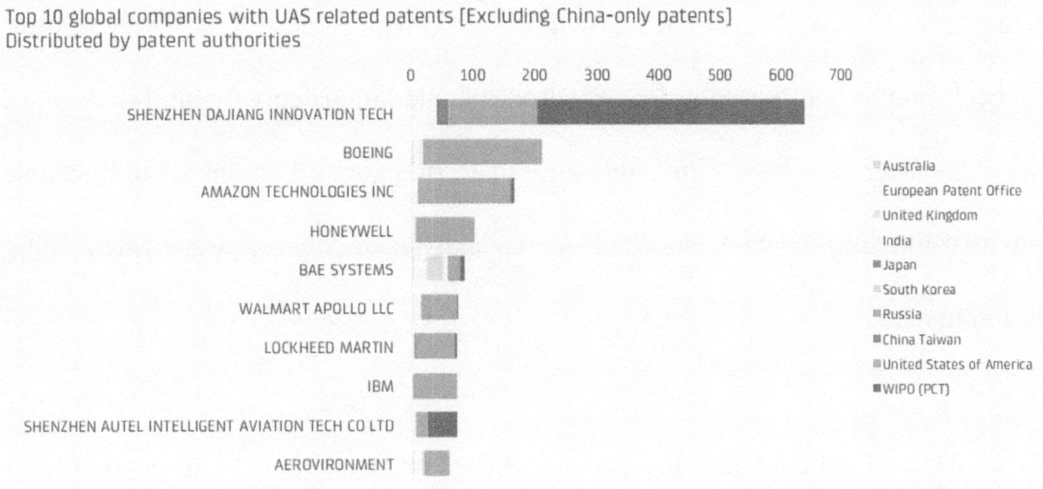

Source: Calculations by Danish Technological Institute based on a search in global patent databases via PatSnap. **7.689 patent families** identified. Syntax: TAC:("Unman*" $w1 "Aircraft*") OR TAC:("remote*" $W1 "aircraft*") OR TAC:("Micro Air Vehicle*") or TAC:("Unman*" $w1 "Aerial*")) not ((AUTHORITY:CN)). Extract 7. dec. 2018. All patents – except patents granted in China only.

The Boeing Company holds the second largest number of drone-related patents. Its UAS projects are used for intelligence, surveillance, and reconnaissance. These include the Phantom Eye, a liquid hydrogen-fueled, high-altitude and long-endurance unmanned aircraft system; and the MQ-25, designed with a robust refueling capability for Navy operations. Boeing has also pioneered innovation in UAS safety. With more drones and UAVs in the air, there are concerns vehicles will collide. In 2016, Boeing was granted a patent for automatic overrides – an important step towards the safe commercial use of drones. However, Boeing is not the only aviation company expanding aerial innovation to drone technology. Airbus, Israel Aerospace Industries, and Wing Aviation also rank among the top 20 patent filing companies. [4]

Amazon, the third ranking non-Chinese patent filing company, began hiring patent lawyers in 2014 to work exclusively in its PrimeAir division. Based on an analysis of patent records, as of August 2017, Amazon has been awarded at least 64 patents for concepts and technologies for delivery drones, including patents for aircraft designs, safety and security systems, methods for transferring goods from the air to the ground, and hive-like fulfilment centers. An in-depth analysis of Amazon drone patent filings can be found here: [5].

Honeywell, a UAS company, has patents focused on its safety UAS inspection service, end-to-end drone-based inspections, and data analytics for industrial companies around the world.

European Patent Holders in UAS

Within Europe, Britain and Germany are the market leaders for filing patents. BAE Systems is the dominant player, followed by Airbus and Saab. See figure 2.12 below.

FIGURE 2.12 NUMBER OF PATENTS BY COMPANIES

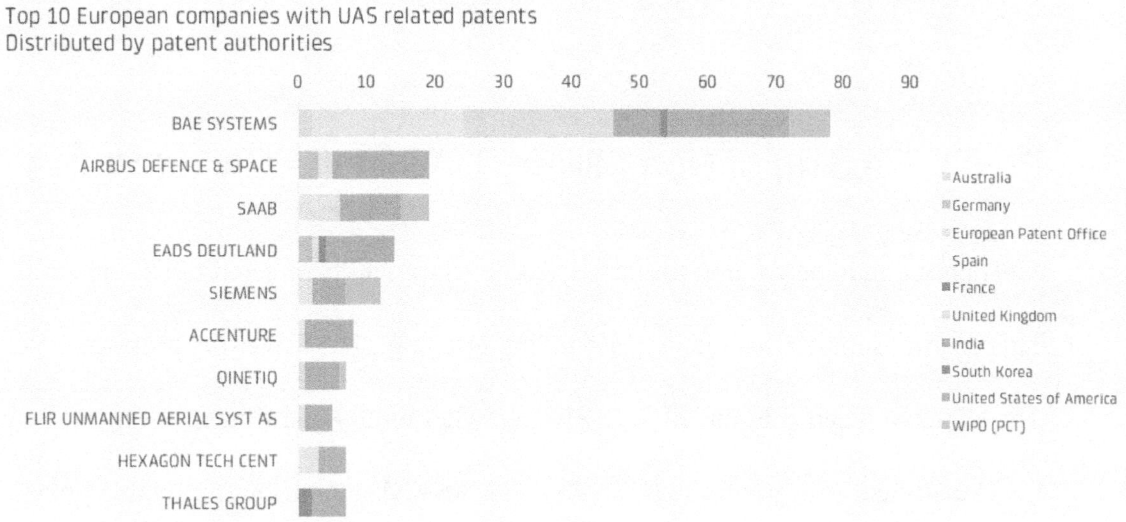

Source: Calculations by Danish Technological Institute based on a search in global patent databases via PatSnap. **1.627 patent families** identified. Syntax: TAC:("Unman*" $w1 "Aircraft*") OR TAC:("remote*" $W1 "aircraft*") OR TAC:("Micro Air Vehicle*") or TAC:("Unman*" $w1 "Aerial*"). Extract 7. Dec. 2018. Only patents from European companies.

Government Grants and Research Institutions

Public Funding of Non-Military UAS in the United States

Most public funding for non-military UAS development is directed to US universities. This chapter summarizes the largest funding grants for UAS over the previous 5 years. The National Science Foundation (NSF) overwhelmingly contributed the most, with the US Department of Agriculture (USDA) and NASA also contributing sizable grants towards UAS research.

Northeastern University opened a state-of-the-art research facility for cyber-security testing on drones. The Air Force Life Cycle Management Center provided a $2.8 million grant to fund research through its unit at Hanscom Air Force Base to enable Northeastern and other industry partners to assess and develop new drone technologies emerging from the commercial marketplace. [6]

The US Army granted $130,000 to Kent State's College of Aeronautics and Engineering to study propulsion systems for a new generation of intermediate-sized unmanned aircraft systems. [7]

The U.S. Department of Defense awarded a $7.2 million grant and a $7.7 million contract to the University of Missouri-Kansas City in 2018 for work on reducing national security threats posed by small, cheap aerial vehicles. Research focuses on developing systems that use high-power microwaves that destroy the electronics of threatening drones. [8]

The NSF awarded Sinclair Community College more than $545,000 to support entrepreneurship and training in drone technology. The funding will be used in three main ways: stoking interest among K-12 students in UAS and STEM fields, creating a new one-year certificate in UAS entrepreneurial training, and creating a UAS-focused Kickstarter competition. [9]

Louisiana's Houma-Terrebonne Airport received a $1.3 million federal government grant by the Department of Commerce's Economic Development Administration to advance its unmanned aviation system. The airport commission has been working to develop it's modern aircraft system through public-private partnerships as a way to diversify and support the local economy. The program, which requires a $337,500 local match, is in conjunction with Nicholls State University, Terrebonne Economic Development Authority, Fletcher Technical Community College, and other private companies and is called the Gulf of Mexico UAS Center of Excellence, or UGC. The UAS program plans to begin remotely operating cargo operations by 2020, with fully autonomous aircraft coming to the region by 2025. [10]

Four companies have each received a part of a total $6 million in funding for UAV research and development from the Ohio Federal Research Network. A Dayton-area company, Surveillance Systems, won an award for autonomous/remote piloted "Air

Uber" system. The other three projects include the Regional Unmanned Traffic Management System led by University of Cincinnati; UAV Detect-and-Avoid Sensor Fusion, led by Ghostwave Inc. in Columbus; and a Brushless-Doubly-Fed Machine (BDFM) and Drive System led by The Ohio State University. Projects must focus on priority research initiatives of the Air Force Research Lab, Naval Medical Research Unit Dayton, National Air and Space Intelligence Center, and NASA's Glenn Research Center. Each project includes at least two Ohio universities, one industry member, and engagement with an Ohio-based arm of a federal partner. [11]

The NSF granted $35 million in research funding between 2016-2021 "to accelerate the understanding of how to intelligently and effectively design, control, and apply UAS to beneficial applications." This includes areas such as monitoring and inspection of physical infrastructure, smart disaster response, agricultural monitoring, the study of severe storms, and more. [12]

N.C. State University and several partners gained a $24 million grant from the NSF and U.S. ignite to create the first wireless platform to link drones and emerging next-generation 5G technology. U.S. ignite is an organization seeking to drive broadband availability and use across the country. [13]

SMU's Lyle School of Engineering won a $849,839 grant from the NSF to improve UAS communications. Research could enable the next wave of drone applications ranging from delivery of consumer goods to supporting autonomous combat and search and rescue efforts. The objective is to build infrastructure for Multi-Dimensional Drone Communications Infrastructure (MuDDI) to address research issues related to three-dimensional connectivity, distributed antennas across a drone swarm, and 3-D swarm formations that optimize the transmission to intended receivers. [14]

Colgate University received a five-year $500,000 NSF grant to study permafrost thaw in Antarctica and create a new integrated education plan for training student research drone pilots. It enables a geologist to take four Colgate students to Antarctica to assist with mapping wetlands, and to work with another six in the lab, in an effort to better understand how thawing permafrost impacts carbon sequestration and resulting mineral runoff in surrounding oceans. To do this, they are planning to take an innovative approach of attaching spectrometers to drones to measure soil saturation. [15]

The NSF granted $6 million in 2015 to fund the four-year Collaboration Leading Operational UAS Development for Meteorology and Atmospheric Physics (CLOUD-MAP) project. Oklahoma State University, the University of Oklahoma, the University of Kentucky, and the University of Nebraska will work together to develop the capabilities of meteorologists and atmospheric scientists to use unmanned aircraft as an everyday research tool. [16]

The NSF awarded a $1.49 million grant in 2019, part of its National Robotics Initiative 2.0 program, to Oklahoma State University. The grant funds research that will use drone technology to estimate wind conditions in differing environments, as well as investigate how to improve the safety and efficiency of drone operations based on wind estimates.

[17]

The NSF awarded a $1 million grant for multiple drone research and development projects to be divided between the University of North Texas, the University of Texas at Arlington, Texas A&M University – Corpus Christi, and the University of Puerto Rico at Mayagüez. [18]

The University of Cincinnati's College of Engineering and Applied Science got a share of a $6.3 million federal grant to continue developing UAVs for traffic monitoring and construction management projects. UC has a partnership to help Ohio's Department of Transportation create new UAV tools for bridge inspections, traffic counts, and construction oversight. UC's regional unmanned traffic management system was one of four projects selected by the Ohio Federal Research Network, a collaboration created by Ohio lawmakers in 2015 to generate business opportunities in Ohio. The goal of the program is to spur innovation in Ohio's aerospace industry. The grant was awarded under the group's initiative to sustain Ohio's defense and aeronautical readiness by helping institutions and industry overcome critical technical barriers. The other selected projects include a study of autonomous flying cars, detect-and-avoid navigation sensors, and a brushless doubly-fed machine and drive system. The awards were granted to projects that served the research priorities of the Air Force Research Lab, the Naval Medical Research Unit Dayton, the National Air and Space Intelligence Center, and NASA's Glenn Research Center. [19]

Georgia State was awarded a $1.2 million federal grant to develop the use of drones in wildfire management, providing more timely data that could allow authorities to give residents in affected areas more time to evacuate, as well as helping firefighters working on the ground. The four-year grant was awarded by the US Department of Agriculture-National Institute of Food and Agriculture through the National Robotics Initiative 2.0: Ubiquitous Collaborative Robots program. [20]

The NSF's Cyber-Physical Systems program, co-sponsored by The U.S. Department of Agriculture's National Institute of Food and Agriculture, awarded $843,000 to a team at Penn State's College of Agricultural Sciences. The three-year project is aimed at reducing the risk of crop damage by using UAVs to monitor air temperatures on cold nights when there is frost and sending commands to ground robots with heaters mounted on them. In this way, growers would target only areas most at risk - minimizing energy use - and ensure that all parts of the orchard are protected. [21]

The Department of Sociology of the University of Central Florida and collaborators at the Smithsonian Institution, Cornell University, and the University of California-Davis received a $1.25 million dollar grant in 2018 from the NSF to look at eelgrass health along North America's Pacific coast. They will use drones to provide high-resolution mapping and drone imagery of eelgrass communities along the coast. GIS, or Geographic Information Systems, is a framework in which data of spatial locations can be collected, managed, and analyzed to create layered maps and other visualizations. The sampling will quantify changes in eelgrass extent and habitat quality at a much higher resolution than most satellites can provide. [22]

Washington State University's School of Mechanical and Materials Engineering received a $1.8 million grant from the U.S. Army in 2019 to demonstrate a liquid hydrogen-powered UAV and refueling system. The $7.2 million total grant includes researchers from Mississippi State University, Insitu Inc., and Navmar Applied Sciences Corporation. [23]

The NSF granted $500k in 2018 to the New Jersey Institute of Technology for its SoarNet (free Space Optics as bAckhaul and energizeR for drone-assisted NETworking) program. SoarNet aims to simultaneously and rapidly transmit data and energy from an access node to a drone-mounted Base Station (DBS), which can be flexibly deployed to provision wireless broadband access. That is, the DBS can simultaneously receive high-speed data streams and energy via optical beams. The received energy is used to power the DBS to prolong its flight and received data streams are delivered to Mobile Users (MUs) via existing Radio Frequency (RF) channels. SoarNet will be a game changer for mobile access. By leveraging Free Space Optics (FSO) communications and the drone-assisted mobile networking framework, SoarNet will significantly enhance the throughput of the network and Quality of Service (QoS) of MUs and will advance the state of the art of wireless networking and wireless charging. Proposed research activities will advance the understanding of simultaneously charging the DBS and transmitting data at high speed. [24]

A $500,000 research grant in 2019 was awarded by the U.S. Department of Agriculture

(USDA) to Embry-Riddle Aeronautical University to combine state-of-the-art drone-captured imagery, meteorological data, and Light Detection and Ranging (LiDAR) data with commonly available Geographic Information Systems (GIS) data and satellite imagery. [25]

The NSF awarded $500k to the University of Southern California in 2018 to develop a concept for a coordination system that is able to find collision-free paths for a large number of flying unmanned air vehicles of different size and capability. This project provides prototype algorithms for managing this dense drone traffic. It uses a hierarchical approach, combining centralized and local coordination, to manage complexity for a large-scale problem. This approach can scale up to handle thousands of drones and lead to conflict free flight. It demonstrates the concept using mixed-reality simulations and using existing helicopter-like robots on a smaller scale. [26]

The Department of Agriculture (USDA) awarded Purdue University $703k in 2017 to collect images of soybean yield plots via drones for three seasons and develop analytical tools to quantify parameters from them that are useful for soybean breeding and production. The objectives are to create software tools for imaged-based proximal remote sensing of crop canopies, establish new phenotyping methods using proximal remote sensing for traits, and demonstrate canopy coverage selection in the early generation stages of soybean breeding pipelines using high-throughput UAS platforms. [27]

The NSF in 2019 awarded $1.2M to the University of Notre Dame for the novel DroneResponse platform. This platform represents the next generation of emergency response solutions in which semi-autonomous and self-coordinating cohorts of drones will serve as fully-fledged members of an emergency response team. Achieving the DroneResponse vision involves delivering novel scene recognition algorithms capable of recreating high-fidelity models of the environment under less than ideal environmental conditions. The work addresses non-trivial cyber-physical systems (CPS) research challenges associated with (1) scene recognition, including image merging, dealing with uncertainty, and geolocating objects; (2) exploring, designing, and evaluating human-CPS interfaces that provide situational awareness and empower users to define missions and communicate current mission objectives and achievements, (3) developing algorithms to support drone autonomy and runtime adaptation with respect to mission goals established by humans, (4) developing a framework for coordinating image recognition algorithms with real-time drone command and control, and finally (5) evaluating DroneResponse in real-world scenarios. Researchers will leverage user-centered design principles to develop human-CPS interfaces that support situational awareness designed to enable emergency responders to make informed decisions. The end goal is to empower human operators and UAS to work collaboratively to save lives, minimize property damage, gather critical information, and contribute to the success of a mission across diverse emergency scenarios. [28]

The NSF in 2016 granted $602k to Northern Arizona University to develop a new UAV system that integrates a radio telemetry receiver and data processing system for efficient detection and localization of tiny wildlife radio telemetry tags. The technology will benefit both scientific research and society by quickly improving the ability to understand the geographic distributions and habitat needs of many animal species. Current methods of locating and tracking small tagged animals are hampered by the inaccessibility of their habitats. The high costs, risk to human safety, and small sample sizes resulting from current radio telemetry methods limit the understanding of the movement and behaviors of many species. UAV--based technologies promise to revolutionize a range of ecological field study paradigms due to the ability of a sensing platform to fly in close proximity through rough terrain at very low cost. The complete design, including parts, assembly plans, software, and training modules for the UAV--RT system will be open sourced for widespread adoption of the technology, with dissemination and outreach by a user community website and demonstrations at wildlife conferences. A team of engineers and wildlife ecologists from Northern Arizona University and Cornell University will design and build the prototype system and conduct field deployments to demonstrate and improve its performance. [29]

NASA granted Texas A&M University $717k in 2019 to integrate terrain analysis and science operations to improve operational and scientific outcomes during exploration of mafic detrital environments. The objective is to evaluate the effectiveness and capability of rover-based semi-automated terrain analysis (using the Automated Soil Assessment System, ASAS) and UAS for navigating, selecting targets for sampling, and characterizing mafic detrital sediments along a glacio-fluvial-aeolian sand transport pathway in Iceland. [30]

NASA issued a $795k grant in 2015 to University of Kansas Center for Research Inc. to develop and demonstrate a multichannel radar to provide small UAS with situation awareness regarding the relative position and velocities of nearby objects (both mobile and fixed) that may pose a collision risk. Objectives include development, testing, and flight evaluation of a proof-of-concept multichannel radar brassboard on a Cessna-172 aircraft (owned and operated by the Kansas University Aerospace Engineering (AE) department) for performance evaluation. [31]

Virginia's government drone initiative received a $1.5 million grant to develop a research institute dedicated to autonomous systems from the Virginia Research Investment. The three-year grant will fund the Commonwealth Center of Innovation for Autonomous Systems (C2IAS). The center is a collaboration between Virginia Tech and Old Dominion University. Virginia Tech is also a participant in the UAS Integration Pilot Program. Private companies will collaborate and invest more than $2.2 million in matching funds. Companies include UAVPro, Atlas-NA, Burton-LeRoy, Cambrian Design and Development, Procomm Aviation, Sanchez Engineering Services, SimIS, and Textron. The center will develop unmanned systems for first responders, infrastructure inspection, maritime systems and more. Commonwealth leaders continue to promote drone careers as an alternative to the dying coal industry. [32]

UAS Startups & University-Private Company Joint Ventures in UAV

At least 70 different companies are currently working on drone defense technologies after having raised over $2B from 2010 to 2018 across the industry. Some examples are $20M raised by Airspace Systems, $12M raised by Citadel Defense, $15M by Dedrone, and $11.5M by SkySafe.

GRYFN, an agbioscience startup, partnered with Purdue University in 2019 on a $4.5 million grant from the Advanced Research Projects Agency-Energy (ARPA-E) to create precise geomatics solutions for coaligned and repeatable multi-sensor drone data collection. The approach enables breeders to scale research operations for precise, repeatable analytic solutions of high throughput phenotyping in the field. [33]

Drone-inspection company SkySpecs raised $29 million in 2019 from McRock InFund L.P. of Toronto, Equinor Energy Ventures of Norway, and Evergy Ventures of Kansas City. Follow-on investors include Statkraft Ventures of Germany; UL Ventures of Northbrook, Ill.; Capital Midwest Fund of Mequon, Wis; and First Step Fund of Detroit. The company also won a small-business innovation research grant of $150,000 from the NSF. Their autonomous drones dominate the autonomous wind-turbine inspection market, with customers in 19+ countries on all five continents. SkySpecs makes sensors and software to enhance drones produced by other companies. [34]

Microsoft awarded a $100,000 TechSpark grant to support Airtonomy, a startup that's partnering with the University of North Dakota Aerospace Foundation for drone applications in North Dakota's "Silidrone Valley." The seed money unlocks nearly $570,000 in additional funding for Airtonomy from local investors. Airtonomy combines drone technology and artificial intelligence to help clients in agriculture, energy, and public safety to realize the benefits of aerial imagery provided by multi-drone systems. [35]

Impossible Aerospace received a $9.4 million Series A from Bessemer Venture Partners, Eclipse Ventures, and Airbus Ventures to develop it's 'all battery' US-1 drone that can stay airborne for about 120 minutes in optimal flying conditions, with a 75km (over 46 miles) straight-line range. It can carry 2.9 pounds of payload, but that drops the total flight time to 78 minutes. This far exceeds the typical 20-35 minute flight times of most small commercial drones. [36]

SkySafe raised an $11.5 million Series A funding in 2017, adding to $3M in seed funding from the previous year, to develop it's radio wave technology that can detect and stop rogue drones from entering unauthorized areas like military bases, stadiums, prisons, and airports. SkySafe's radio frequency signals are projected from a perimeter of nodes or even a Jeep, and force unapproved drones to leave or land while allowing permitted drones to fly. SkySafe's mobile defense vehicle can accompany armed forces in the field to protect a moving perimeter from drone attacks or surveillance. Meanwhile, SkySafe has secured a $1.5 million Department of Defense contract with Naval Special Warfare to provide counter-drone tech to the Navy Seals. [37]

WhiteFox received a $12M extension of seed funding from investors like JAM Capital, Stage Venture Partners, Okapi Venture Capital, Serra Ventures and OCA Ventures for expansion. It has developed technology that's based on identifying and differentiating between drones based on their unique radio frequency signatures. That product for identifying drones is complemented by a second technology offering that allows WhiteFox to take control of the unauthorized drones in an airspace. [38]

Military Funding and Research

Currently, military spending accounts for more than half of the entire drone market, according to research by global tech market advisory, ABI Research.

The Center for the Study of the Drone at Bard College published two comprehensive reports on US military drone spending, one on the 2018 Defense Budget and the other covering the 2019 proposed Defense Budget. Much of the following summarized information and charts below were derived from these reports. You can download them in full here: [39][40]

OVERALL DOD FUNDING FOR UAV'S

The US Department of Defense Fiscal Year 2018 budget request contains $6.97 billion for drone-related procurement, research and development, and system-specific construction. The Pentagon's annual summary of funding for major defense systems notes that $2.6 billion has been allocated to existing unmanned aircraft in FY 2018, not including procurement of spare parts for existing drone systems, unlisted items within larger projects, and classified systems not available to the public. The single largest item in the DoD request is the MQ-9 Reaper, which is allocated $1.23 billion in the FY 2018 request.

There was an approximate +6.6% budget increase per year from 2013 to 2018. Public budget data indicates that funding for drones in Fiscal Year 2018 was significantly greater than what the Pentagon predicted in 2014. The ups and downs of the drone budget since 2013 have generally tracked with the overall military budget. See below figure 3.1.

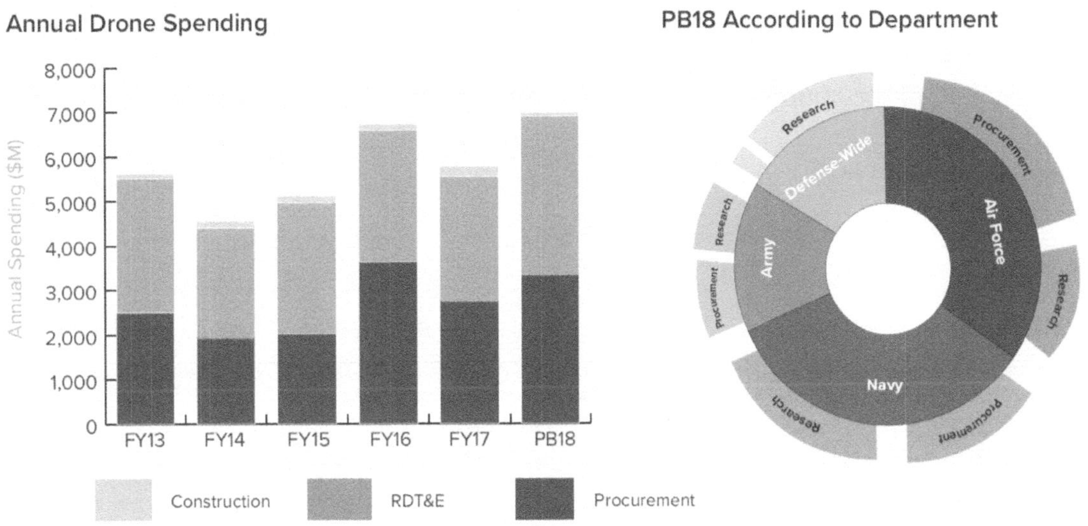

Figure 3.1

Between FY 2013 and PB18 the military spent on average 79% of the drone budget specifically on air systems. Of that, 66% was on legacy aircraft and 13% was on UAS capabilities and future systems. In PB18, research and development spending was 51 percent of the drone budget, while procurement was at 48 percent.

The DoD roadmap recommends that future systems should be "low cost, disposable platforms" that are better suited to contested environments compared to expensive, multi-role drones like the Reaper and Gray Eagle.

For existing unmanned aircraft, budgets are now focused on modernization efforts rather than acquisition. In 2012, the military purchased 1,232 drones. Most of these - 1,140 - were hand-launched RQ-11 Ravens. But the military also ordered 48 MQ-9 Reapers, 29 MQ-1C Gray Eagles, 12 MQ-8 Fire Scouts, and three RQ-4 Global Hawks. Since then orders for unmanned aircraft have dropped off. In PB18, the military requested funds to purchase 30 of these aircraft and four RQ-21A systems. Figure 3.2 below shows a system level spending breakdown.

System-Specific Spending

Category	PB18
MQ-9 Reaper	1,234.8
RQ-4 Global Hawk and AGS	427.9
MQ-8 Fire Scout	164.6
MQ-4 Triton	988.2
RQ-21A Blackjack	115.8
MQ-1C Gray Eagle	358
RQ-7 Shadow	126.4
MK 18 Family UUV	63.1
MTRS II Robot	53.5

Drone RDT&E Budget Activity Breakdown

Budget Activity	PB18
BA 1, Basic Research	66.7
BA 2, Applied Research	478.4
BA 3, Advanced Technology Development	452
BA 4, Demonstration and Validation	928.7
BA 5, Engineering and Manufacturing Development	403.3
BA 6, RDT&E Management Support	8.1
BA 7, Operational System Development	1,217.6

Figure 3.2

In FY 2015 and FY 2016, DARPA issued 77 companies and institutions contracts for 16 different drone-related research programs as listed in figure 3.3.

Company/Institution	Total ($M)
Aurora Flight Sciences	81
Lockheed Martin	55.7
Northrop Grumman	53.4
Raytheon	51.1
Leidos	34.8
Boeing	15.2
Sikorsky	14.7
AeroVironment	13
Charles Stark Draper Laboratory	11.6
Applied Physical Sciences Laboratory	10.6
BAE Systems	10.1
Scientific Systems Company	9.4
Johns Hopkins University Applied Physics Laboratory	8.7
L-3 Communications	7.7
Six3 Advanced Systems	6.7

Figure 3.3

Air Force Funding Plans

The Air Force "SUAS Flight Plan 2016-2036" report notes that many of the capabilities needed today can be accomplished with much smaller and cheaper aircraft than the very large aircraft of the past. In PB18, the Air Force allocated an unspecified amount to support the development of "attritable" aircraft (small, low-cost armed drones). However, the long-term viability of legacy small, hand-launched drones also faces uncertainty when affordable consumer systems can be adapted for military use for significantly lower cost than most military systems. Some of these systems, like the AeroVironment Raven, which costs around $35,000 each, are viewed as too expensive and bulky for infantry to carry around. The Army's Soldier Borne Sensor program is one effort being made to come up with an alternative. In PB18, the Army allocated $3 million to purchase 300 Soldier Borne Sensors. Although the Army has not selected what type of drones it will purchase as part of this program, they could be something like the DJI Spark drone, which can be easily stored and carried. The Pentagon also allocated $650,000 to acquire 450 quadrotor drones for U.S. allies in the counter-ISIS campaign.

The Missile Defense Agency had two active drone-related items in PB18, both involving the Agency's effort to develop a laser-equipped drone capable of shooting down ballistic missiles.

The Office of the Secretary of Defense's (OSD) AVATAR program allocated $25 million for a project that would enable fighter pilots to control converted manned aircraft and target drones. DARPA's Gremlins, allocated $37.8 million in PB18, is developing small reusable drones that can be deployed en masse for intelligence collection. Squad X, a DARPA project allocated $36 million in PB18, is working to integrate unmanned air and ground vehicles into small infantry units. Collaborative Operations in Denied Environments (CODE), allocated $30 million in PB18, is developing algorithms and software that will enable drones to operate collaboratively, autonomously, and with relatively little oversight by nearby fighter pilots.

The thrust of the big DARPA and OSD-funded research projects in PB18 reflects the belief that the next generation of unmanned systems will need to be smaller, faster, and, perhaps most importantly able to work collaboratively with humans and other drones. This view is not restricted to OSD and DARPA. The Air Force is developing low-cost attritable drones - armed unmanned aircraft that could accompany and protect a manned fighter jet.

The Department of Defense requested approximately $6.05 billion for unmanned aircraft systems in the FY 2019 budget. The FY 2019 budget proposal represents a significant expansion in drone spending over the FY 2018 requested budget, up from around $5 billion in both the FY 2018 request and the FY 2017 appropriated budget. The aircraft budget is divided between $3.71 billion in procurement, $2.14 billion in research and development, and $198 million in military construction. The DoD intends to purchase a total of 3,250 unmanned air vehicles. The FY 2019 request also contained orders for 1,618 AeroVironment Switchblade loitering munitions - the largest to date - and 532 other unmanned systems such as aerostats and target drones. Compared to the FY 2018 request, Navy funding for unmanned systems increased by approximately $1 billion (38%) and Army funding increased by $719 million (73%). A more than $500 million boost to the Navy's MQ-25 Stingray research and development program is the single greatest contributor to the overall increase in drone spending.

In the Fiscal Year 2019 proposed budget, the U.S. military is aiming to significantly expand its inventory of small unmanned air systems. After years of budget stagnation for small unmanned aircraft programs, each of the military services is directing fresh funding streams both to "legacy" small drones like the AeroVironment RQ-11 Raven and to new procurement projects. The procurement budget for small drones is around $279 million, up from $89 million in the FY 2018 request and $136 million in the FY 2017 appropriation. These funds will be used to procure approximately 3,070 new small drones.

DoD is looking specifically at drones that can improve the situational awareness of small units and that are able to perform complex missions in urban environments such as reconnoitering a building.

In future years, the Army predicts it will spend around $105 million on a new, smaller type of drone (likely not the Raven) for its Short Range Micro UAV requirement. One Raven system costs $180,400. It has allocated $21.7 million for the Soldier Borne Sensor program to purchase 1,084 systems. The SBS program is designed to give an infantry squad a lightweight drone that can be easily carried and deployed for short-range, quick-look reconnaissance. One SBS is estimated to cost $15,923.

Marine Corp Funding

The U.S. Marine Corps requested $2.9 million for 200 unmanned vertical takeoff and landing air vehicles. These are likely the Physical Sciences InstantEye Mk-2 GEN3-A0 quadrotor, although other aircraft types may be considered. Like the Army's SBS, this program envisions equipping Marine infantry squads with their own nano vertical takeoff and landing drone for short-range reconnaissance. One system costs $28,950.

US Operations Special Command Funding

The U.S. Special Operations Command requested $74.7 million to acquire a variety of unmanned air systems, including $38.9 million for seven different types of small drones. The systems range from quadcopters to Unmanned Tethered Aerial Systems. The largest SOCOM order is $10.5 million for 527 Nano VTOL drones, which could resemble the Marine Corps' InstantEye UAV.

DOD Counter-drone Budget

Funding for counter-drone systems (C-UAS), including a shipboard laser, continues to rise. The DoD's counter-drone budget rose from $528.8 million in the FY 2018 request to $1.05 billion in the FY 2019 proposal—$468.8 million in procurement and $583.9 in research and development. Of this requested funding, $280.8 million is allocated for the Army's Indirect Fire Protection Family of Systems procurement, which will buy 25 counter-drone systems currently known as the Low-slow-small Integrated Defense System (LIDS). The Army has allocated an additional $188.3 million in RDT&E funds for work on LIDS and other C-UAS solutions.

DARPA's two dedicated C-UAS projects are Mobile Force Protection, allocated $37 million, and Aerial Dragnet, allocated $18.2 million. The former project is dedicated to developing a system capable of defending a convoy from a swarm of self-guided drones, while the latter project seeks to develop a system capable of detecting drones in an urban environment.

EU Government Spending

The EU drastically increased its spending on military research in 2019 to approximately €4.1 billion (US$4.6 billion) for the period 2021-27. Previously, the first outlay in research was small: a €1.3-million pilot scheme beginning in 2016, which included funding for EuroSWARM, followed by a three-year, €90-million programme in 2017. The European Parliament's decision to create the European Defence Fund increases annual funding for military research by more than ten times, to around €500 million a year, starting in 2021. And it would also boost annual spending on military industrial development and prototyping to around €1 billion a year. Together, they total an investment of €13 billion from 2021 to 2027.

So far, the pilot phase and initial three-year programme for research have focused mostly on funding technologies that are relatively late in their development. One example is the €35-million OCEAN2020 project, which trialed ways to improve surveillance at sea by integrating drones and crewless submarines into operations involving several navies.

The skew towards later-stage development has been attributed to why academic institutions received less than 2% of the €44 million allocated, whereas 71% went to companies and 26% went to research and technology organizations (RTOs) — such as the Netherlands Organization for Applied Scientific Research (TNO). This contrasts with

Europe's €75-billion Horizon 2020 research programme; from 2014 to 2016, higher-education institutions took 39% of that funding. [41]

Chinese Drone Research and Spending

It has already been demonstrated that China dominates in terms of patents for drones, however the Chinese funding environment is more opaque. This report will attempt to shine a light on the funding and research environment within China using information derived from public English-language publications.

University Research on UVA in China

Universities have been the traditional backbone of China's drive to develop drone technology. They still play a prominent role with several universities enjoying governmental support to develop the field.

Beihang University (Beijing University of Aeronautics and Astronautics) is China's leading aeronautics university with strong links to European institutions with whom they collaborate on faculty mobility, student exchange and research projects, and also has a number of Erasmus+ programs. Website: http://ev.buaa.edu.cn/

North-western Polytechnic University is a highly internationalized university with more than 150 international partners, 18 double degree programs, and 30 short term exchange programs. Their Unmanned Aerial Vehicle Research Institute is among the leading institutes in China and linked to the drone manufacturer and tech company Aisheng. Website: http://www.nwpu.edu.cn/

Nanjing University of Aeronautics and Astronautics is among the leading institutions in China in the field of Unmanned Aerial Vehicles. Website: http://iao.nuaa.edu.cn/

Harbin Institute of Technology (HIT) is one of China's top universities. It does not however have a dedicated school/institute for aeronautics but features schools in transportation, civil engineering, and computer science. HIT collaborates with many international universities. Website: http://en.hit.edu.cn/

Private Industry Research of UVA in China

A report published by Chinese language industry media outlet 81UAV said that by 2025, China's drone industry will be worth $114B, with aerial photography, forestry, agriculture, and security all being key areas for development. [42]

Chinese drone maker Yuneec received $60M from Intel to form a partnership in 2015. It culminated in the creation of Intel's RealSense Obstacle Avoidance, a system for automatic obstacle avoidance for drones. The RealSense Follow Me mode films in all directions to ensure collisions with objects are avoided. The Intel RealSense R200 camera with Intel Atom powered module builds a 3D model of the world, allowing customers to focus on the subject without worrying about flying into obstacles. [43]

In 2017, Chinese drone research and developer Shenzhen MicroMultiCopter Aero Technology Co. secured $30.3 million in its pre-IPO funding round. The company was founded in 2014 after setting up a research team in 2008. It raised $4.09M in A-round funding and $6.1M in B-round funding in 2015 for a valuation of $76M, and $9.1M in 2016 for undisclosed purposes. MicroMultiCopter's unmanned aerial vehicle products are used in such fields as agriculture, public security, fire control, and aerial photography. The firm has set up four research and development centers and cooperated with well-known firms, including China Aerospace Science and Technology Corp. MicroMultiCopter is currently training drone pilots. [42]

Ele.me, one of China's leading online food delivery platforms, was given the green light by Shanghai authorities in 2018 to operate in the country's initial delivery routes for drones. The on-demand service provider, which is owned by Alibaba, has applied for more than 60 patents on drone and driverless car delivery, and plans to achieve mass application of unmanned delivery next year. Under the rules drawn up for this programme in Shanghai, drones will transport meals to and from two fixed locations within each route. The 17 newly approved routes over an industrial zone in Shanghai will enable consumers to receive online takeaway meal orders via drones within 20 minutes after confirming the delivery on their smartphones. [44]

Public Sector Research on UVA Technology in China

In 2018, research was detailed on the EU-China APP program to develop closer cooperation between EASA and the CAAC. The EU-China APP was initiated in September 2015 with a duration of five years (2015-2020) and an EU budget of EUR 10 million to boost the development of China's Civil UAS Aviation Operation Management System (UOMS). The UOMS UTM traffic management system will be integrated within the CAAC's General Aviation Flight Service (GAFS) system and current ATM system. Remotely piloted aircraft systems (RPAS) operations in China are currently supported by three types of ground control stations – take-off, en-route, and landing. Ground stations and current ATM control systems interact through the link relay of the RPAS itself. The RPAS switches over multiple crews during operation. Work underway to progress the development of the UOMS network includes drawing up standard specifications for the data interface between the U-cloud – where drone flight information is stored – and UAS operations, and developing a geo-fencing system which will be built into the flight control system. The geo-fencing system is based on Real Time Kinematics (RTK) high precision-positioning techniques. The three major UTM trials underway are the UAS Traffic Management Information Service System (UTMISS); the UAS cloud systems for small UAS management, unmanned cargo flight, and large UAS; and lastly the information sharing between GAFS and UOMS. [45]

In 2018, Shanghai's Civil Aviation Administration of China and the Chinese Air Force's Shanghai command came together to establish their new Drone Innovation Center in Shanghai. The facility will provide participants with 36 square miles of airspace, and is intended to boost research, development, and testing for industrial drone applications being considered by the state. The three main elements being prioritized are flight testing, demonstrating the intended drone applications, and refining their abilities for eventual industrial implementation. There are also serious considerations of testing water-borne takeoff and landing drone systems. Additionally, the compound will research law enforcement and firefighting applications, and serve as a manufacturing and assembly facility for drone companies in the future. [46]

China's Role in UAV Scientific Publishing

According to the 2017 report, "International Opportunities for Danish Drone Researchers," researchers from the US published 1,110 drone publications in 2016. Followed by China with 866, South Korea with 226, and Germany with 211. The number of drone publications increased by 40% between 2013 and 2016. According to this survey, USA and China accounts for 26 and 20 percent, respectively, of drone publications worldwide. China's focus on drones initially concerned national security and the use of drones for military purposes. However, the commercial potential was quickly realized and so funding for developing the industry was provided not only by the Department of National Defense but also by the Ministry of Science and Technology, and the Ministry of Industry and Information Technology. Thus, today China's commercial drone industry is generally considered to be internationally leading. Compared to their international counterparts, China's commercial drones have a reputation for being less advanced, but they function on a 'need to have'-basis and are generally sold for only 1/5 the price. [47]

China's Military Drones

China has emerged as one of a handful of global leaders in the development of military UAS, producing a range of copy-cat drones very similar to that used by the US military (Fig. 4.1 below). China has carefully factored unmanned systems into its strategic planning as it pushes military modernization and its position as a global arms exporter. It's 5 year plan includes the development of technologies for long-endurance stealth drones and to complete the design of high-speed combat/reconnaissance drones.

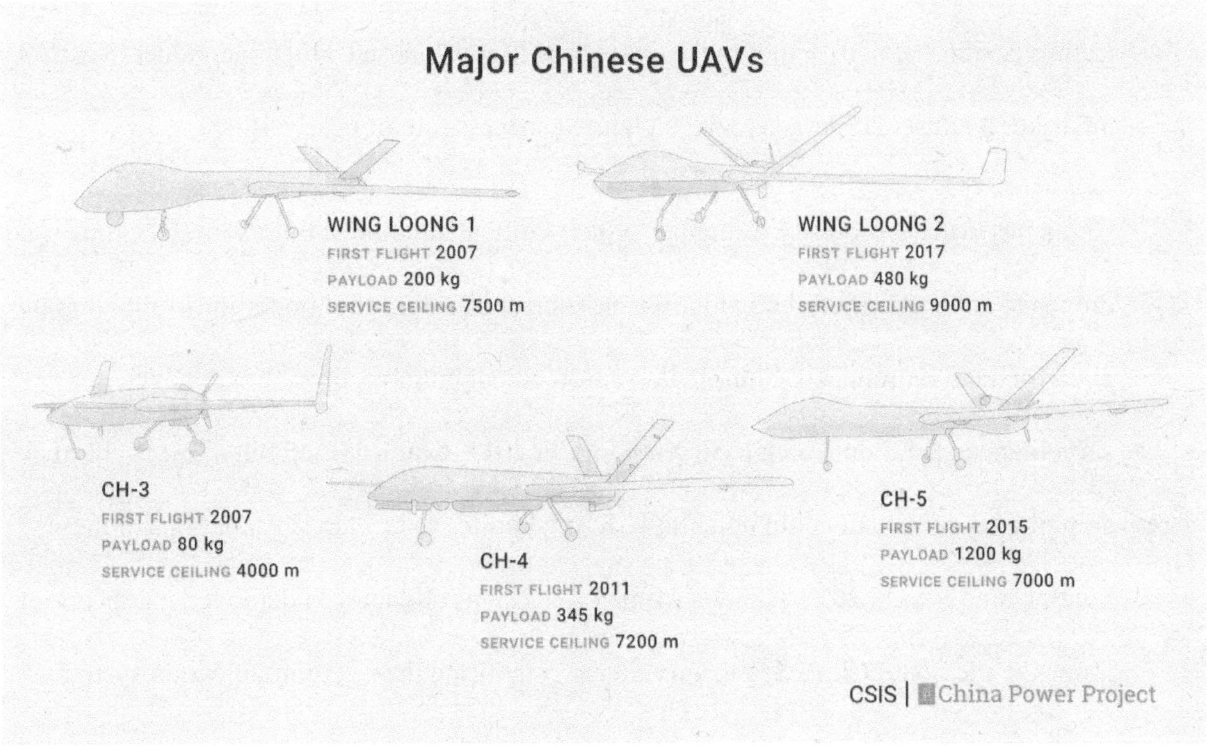

Fig 4.1, source: [48]

President Xi remarked in 2016 that "UAVs are important operational forces for the modern battlefield." China has made considerable progress in establishing itself as a leader in global arms sales, particularly drones, finding success in producing both strike-capable systems and non-weaponized systems for intelligence, surveillance, and reconnaissance (ISR) missions. A 2018 DoD report noted that the People's Liberation Army Air Force (PLAAF) "is closing the gap with the U.S. Air Force across a spectrum of capabilities, gradually eroding longstanding U.S. technical advantages." Its Wing Loong and series drones have become popular exports to militaries around the globe, particularly to the Middle East and North Africa. Its fleet of reconnaissance drones includes the High Altitude Long Endurance (HALE) Soaring Dragon and Cloud Shadow.

There are currently no reports of the Chinese military itself carrying out drone strikes. However, other militaries have used Chinese built drones to carry out attacks. For example, since late 2015, the Iraqi military has used Chinese CH-4s to conduct over 260 airstrikes against ISIS targets, with a claimed success rate of nearly 100%.

China has utilized drones in a number of non-combat situations. For example, China used drones to support various humanitarian assistance/disaster relief operations following the 2008 Sichuan Earthquake. Chinese law enforcement employs drones to conduct surveillance operations, such as in Xinjiang. In 2017, China carried out a test flight of an amphibious drone that could potentially ferry supplies to military installations in the South China Sea. In 2019, China's Ministry of Natural Resources deployed a network of drones to the South China Sea to surveil and establish a drone communication system of air and land-based UAVs.

China's Military Drones in the International Drone Market

According to UAV Global, China has 21 manufacturers of military drones compared to the USA's 20 manufacturers. While US manufacturers still tend to be more technologically advanced compared to their Chinese counterparts, the gap is closing. At the same time, China's cheap and effective manufacturing has enabled Chinese military drone manufacturers to penetrate the international market. For a comparison, between 2008 and 2018 China has sold 181 drones to 13 different countries. Over 75% of these sales were of strike capable systems, including those of the popular Caihong (35% of sales) and Wing Loong (54%) series. However, China lags leading exporters. Within the same time frame, the US led global drone exports with a total of 292 drones sold to partners around the world. Israel followed closely with the sale of 265 UAVs. The majority (88%) of US sales have been composed of ISR drones, ranging from the surveillance drone Scan Eagle up to larger platforms like the Global Hawk. Roughly 70% of Israel's UAV exports have been ISR platforms, including the Searcher and Aerostar.

The biggest draw to Chinese drones is their price tag. The new CH-5 is estimated to cost half that of the $16 million American-made Reaper. The CH-4 (known as the 'AK-47' of drones) and Wing Loong II are even cheaper, estimated to cost between $1 to 2 million – far less than the comparable $4 million Predator. [47]

South Korean Drone Research and Spending

South Korea is the third most dominant drone producing nation, although still far behind China and the US. Detailed below are the more public drone research and private industry activity within the country. The largest funded drone research drive is currently focused on countering the small spy drone threat of North Korea.

Korean Research on UAS

South Korea's strengths toward UAS rely on its research capabilities. When it comes to UAS, it is often that a company will partner up with an institute or a university to work with government entities to co-develop technologies rather than independently developing their own technologies. Combined efforts from companies, research institutions, and universities in R&D of UAS technologies contribute to high-tech large and medium sized drones.

Main UAS players in South Korea can be categorized as universities, research institutes, and companies. For companies, Korea Aerospace Industries leads in larger and military-related UAVs. As for smaller UAVs, UConnsystems leads the advancement of the drones used for monitoring/surveillance, agriculture, and mapping. Lastly, the universities KAIST and UNIST and the research institutions KARI and ETRI are leading with regards to focus on UAS and UAS technologies such as satellite and navigation technologies.

Korean Government Ministries Responsible for Drone Development

Four ministries in South Korea are responsible for drone/space relevant activities: Ministry of Land, Infrastructure and Transport (MOLIT), Ministry of Science, ICT and Future Planning (MSIP), Ministry of Trade Industry and Energy (MOTIE), and Ministry of Strategy and Finance (MOSF). Each ministry is responsible
for tasks such as establishing regulations for different technologies, supporting R&D activities and commercial activities, as well as distributing funding for these activities to the relevant ministries.

South Korea's public organizations such as the police or fire offices are focused on developing UAS for firefighting, surveillance, and monitoring. Private companies are focused on collaborating with the defense ministry and agencies in South Korea for developing military drones and other related technologies such as artificial intelligence, logistics, and maritime/underwater. [47]

DGIST announced in 2019 that a team in the Collaborative Robots Research Center developed a radar system that can detect subminiature drones that are 3km away. As a result of discovering a North Korean drone in Paju in March 2014, South Korea's Ministry of National Defense has adopted a drone detection radar based on an overseas technology. Since last year, the ministry has devoted itself into building a combat system using drones and training specialized personnel by forming a drone unit to strengthen its defense capability. The necessity of enemy surveillance reconnaissance and the early detection of offensive drones has increased in Korea. Due to the lack of indigenous radar detection technology, South Korea is using many radar systems from abroad such as drone detection radars from RADA in Israel and Blighter in the U.K. RADA especially owns a superior radar technology performance that can detect drones that are a maximum of 3km away. The team has been working since 2014 on a drone detection radar system based on a super-resolution algorithm using only Korean technology. While foreign radars have similar performance, it is noteworthy that key hardware components were developed 100% jointly with Korean small and medium-sized companies. [49]

North Korea has been known to use small UAS for recon of its southern neighbor. As a response, the South Korean defense research institute funded a group of roboticists in 2015 at the Korea Advanced Institute of Science and Technology (KAIST) to test out ways of using autonomous unmanned aerial vehicles (UAVs) to locate, intercept, and disable other UAVs. The group is working with a variety of UAVs with different capabilities. These include agile multi-rotor UAVs equipped with nets that can be dropped on enemy UAVs to disable them. The goal of the project, which is still in its early stages, is developing UAVs that could be used as part of an anti-drone defense system, and that could also go on the offensive if necessary. For example, a scenario in which UAVs must take on a rocket-launching enemy vehicle, which is itself guarded by its own UAVs. [50]

Public Agency Drone Research

South Korea has more than 3,000 islands, of which some 480 are inhabited and where many residents have limited access to health care and other public services. Seoul began testing delivery by drone in the country's remote regions in 2019, with the hope of improving residents' quality of life. The project, jointly launched by the interior and safety ministry and Korea Post, among other government agencies, aims to establish a "public drone delivery system" to serve the country's dispersed population. Seoul aims to open 10 drone delivery bases across the country by 2022. [51]

Photogrammetry software developer SimActive announced in 2018 that it enabled the first government drone mapping project in South Korea. Asia Aero Survey (AAS), a SimActive regional partner, used Correlator3D for the public survey delivered to the Korean National Geographic Information Institute (NGII) and the Korean Association of Spatial Information, Surveying & Mapping. Previously conducted using manned aircraft, a drone was preferred due to the difficulty in accessing the area of interest through traditional aerial and ground surveying methods. AAS is also deploying UAS solutions with hyperspectral and LiDAR for other projects in conjunction with SimActive software. [52]

South Korean Army Research on Drones

The South Korean Army has made an acute note of the country's very low birth rate of 0.98 (2018), noting that this demographic cliff will dramatically reduce the number of new army recruits in the near future. It has thus made a push for drones and other automation to keep up with the much larger militaries of both China and North Korea. Research by the Agency for Defense Development (ADD) is heavily focused on developing medium-altitude long-endurance (MALE) primarily recon drones similar to the US-based MQ-9 Reaper with the potential to carry air-to-ground missiles. Miniature drones on par with America's WASP UAV are in the works. They will be deployed to the new "decapitation" special forces unit established last year by the South Korean Army. These micro drones provide intelligence on targets to the air force and ground troops. The SK Army had established a dronebot military research center in January 2018, with a pilot-scale drone combat unit entering service that same year. [53]

Republic of Korea (ROK) President Park Geun-hye stressed the importance of developing cutting-edge unmanned aerial vehicles and platforms to support the ROK's move to a proactive defense posture as well as the more general value of concomitant improvements to information technologies. At present, ROK UAVs are limited to tactical operations in the airspace above the Korean Peninsula, but the ROK is hoping to deploy UAVs capable of medium-altitude strategic ISR missions in the near future. The ROK's program of UAV development makes use of advanced technologies, many of which are dual use. These include structural materials with stealth technology, avionics control equipment, ultra-high-capacity two-way Long Term Evolution communications technologies, active electronically scanned array (AESA) radar, high-resolution optical sensors, and command and control (C2) systems with cutting-edge ISR capabilities to determine the identification of friend or foe. Individual innovations in these fields are driven by commercial imperatives, which allow ROK MND to adopt these UAV-related technologies to the mutual benefit of both industry and national defense. [54]

South Korea's Defense Acquisition Program Administration (DAPA) in 2019 planned to invest $74 million to develop a weapons system by 2023 that can detect and strike drones after incidents of infiltration by North Korean spy drones. The system, nicknamed Block-I, is designed to track and destroy small drones and other aircraft by locking invisible optical fiber razors on a target at close range. The eventual goal of the project is to develop a system that could also intercept a fighter jet and satellite. [55]

South Korea's current drone detection radar is capable of detecting small drones at a range of approximately five kilometers. It does not have the ability to intercept said drones. [56]

Japanese Drone Research and Spending

Japan is a very minor player in terms of drone funding and research, however it's very advanced technological business environment and strategic importance to the US in terms of conflict with China and North Korea warrants some mention. Like South Korea, Japan's quickly declining population has researchers and officials looking for ways to replace their aging workforce with automation. Our following research includes just a few of Japan's more significant examples.

Japanese University and Public/Private Research

Yaizu City's Crisis Management Division are experimenting with drones that can rescue swimmers, assist in the laying of utility cables and Sony group's drone service that checks for fraudulent building practices, and courier delivery services. Other areas in which Yaizu is utilizing its fleet of four drones includes waterways and farmland management. Other projects relate to the accurate and safe landing of drones on specially developed landing ports and the use of stereo fisheye lenses to allow UAVs to operate indoors and other covered areas where GPS signals struggle to penetrate. [57]

The Autonomous Control Systems Laboratory Ltd. (ACSL) of UTEC, located in Chiba, Japan, is a spinout of Chiba University. ACSL provides industrial unmanned IoT systems, utilizing drones in the area of infrastructure inspection, logistics, delivery, disaster prevention, etc., for the creation and support of Japan's local drone industry. Website: https://www.ut-ec.co.jp/english/portfolio/autonomous-control-systems-laboratory-ltd-acsl

Japanese Private Industry Drone Research

The Japan MAFF (Ministry of Agriculture, Forestry, and Fisheries) said in 2022 drones will account for 50% of spraying solutions for rice planting in Japan. [58]

Japanese drone company Nileworks raised ¥800 million ($7.12 million) to expand the capabilities of its automated crop sprayer drones. The financing round includes a group of Japanese investors including public-private partnership the Innovation Network Corporation of Japan, agricultural chemical maker Kumiai Chemical Industry Co., Sumitomo Corporation and its subsidiary Sumitomo Chemical Co., the Japanese National Federation of Agricultural Co-operative Associations, and The Norinchukin Bank. The company plans to sell the hardware itself, as the drones are automated and do not require operators. Each drone can carry up to one liter of pesticide at a time and fly for 20 minutes, covering 1 hectare. [59]

Drone Fund is a VC fund dedicated exclusively to exploring and enabling unmanned vehicle solutions by investing in companies specializing in hardware, software, DaaS (drone as a service), SaaS, and drone-enabling technology solutions. Drone Fund has 30 portfolio companies, each of which has a specialized and distinct solution. Drone Fund I raised a total of $15 million in January 2018, while Drone Fund II raised a total of nearly 50 million USD in April 2019.

Companies are developing everything from underwater drones, an automated operation drone port in Fujita, UAS powered "smart farming" solutions, a manned hoverbike, a taxi drone system, and a system that utilizes small sized machines that measure atmospheric conditions. [58]

Additional information on Japan's commercial drone industry can be found in the report, "THE DRONE MARKET IN JAPAN," published in 2016: [60]

Japanese Military Spending on Drones

Drones for the Air Self-Defense Force (ASDF) have been developed for the purpose of strengthening intelligence, surveillance, and reconnaissance capabilities. The TRDI and ASDF have been researching drone test systems for quite some time. The procurement of drone systems has been supported by research that has been underway since 2013. The Mid-term Defense Buildup Program (MDBP) from FY2014 to FY2018 has stated that long-endurance drone systems for aerial operations will be used to strengthen all-around intelligence, surveillance, and reconnaissance (ISR) capabilities in a wider area than before and that they will be used in joint operations. These capabilities are needed to reduce psychological and physical crew fatigue during lengthy MSDF and ASDF operations in the strategic environment surrounding Japan. Consequently, the Global Hawk system, a high-altitude and high-endurance drone system, was chosen as a component of Japan's UAV portfolio, with about 15.4 billion yen budgeted for the system in FY2015. [61]

About the Author

John Palmisano has 19 years of experience in all things autonomous systems. Graduated in 2004 from Carnegie Mellon University with a BS in mechanical engineering, he continued on to work a decade at the Naval Research Laboratory (NRL) in Washington, DC, as a robotics specialist and research engineer. There he contributed to robotic systems ranging from creating bio-mimetic underwater drones to designing and building the entire fuel cell control system for the Navy's Ion Tiger drone. During his time at the NRL, he authored and co-authored nearly two dozen scientific papers: https://scholar.google.com/citations?&user=9lhrrPIAAAAJ

In 2016 he moved on to be an independent freelancer doing computer simulations, performing computer based modeling for engineering and cryptocurrency problems, designing and constructing off-grid sustainable systems, and building mass-market commercial electronic systems for clients globally.

Bibliography

[1] Arthur Holland Michel, "Report: Counter-Drone Systems," February 20, 2018.

https://dronecenter.bard.edu/counter-drone-systems/

[2] J Dutczak, "Liquefied and chemical hydrogen storage in contemporary small drones' fuel cell propulsion systems," Materials Science and Engineering, Volume 421, Issue 4, September 2018.

https://iopscience.iop.org/article/10.1088/1757-899X/421/4/042015

[3] "Global Trends - Explosion in the Number of Patents Related to Unmanned Aerial Systems," Danish Technological Institute, Global Trends, 2019

https://www.teknologisk.dk/_/media/74179_Global%20trends%20-%20Explosions%20in%20number%20of%20patens%20related%20to%20unmanned%20aerial%20systems.pdf

[4] Javier Diez-Aguirre, "Flying high: who's leading innovation in drone technology," March 11, 2019.

https://www.cpaglobal.com/cpa-global-blog/flying-high-whos-leading-innovation-in-drone-technology

[5] Arthur Holland Michel, "Amazons Drone Patents", September, 2017.

https://dronecenter.bard.edu/projects/other-projects/amazon-drone-patents/

[6] Ian Thomsen, "State-of-the-art drone testing facility opens on the Burlington Campus," April 8, 2019.

https://news.northeastern.edu/2019/04/08/northeastern-university-in-partnership-with-the-navy-and-air-force-opens-state-of-the-art-drone-testing-facility-on-its-burlington-campus/

[7] Dan Pompili, "Army Funds Kent State Aeronautics Professor to Research Drone Technology," January 11, 2019.

https://www.kent.edu/kent/news/army-funds-kent-state-aeronautics-professor-research-drone-technology

[8] Brian Kaberline, "UMKC gets nearly $15M to swat drones from the sky," Kansas City Business Journal, March 5, 2018.

https://www.bizjournals.com/kansascity/news/2018/03/05/umkc-gets-14m-to-swat-drones-from-the-sky.html

[9] Thomas Gnau, "Sinclair hails $545K National Science Foundation grant for drone work," August 1, 2019.

https://www.daytondailynews.com/business/drone-technology-research-dayton-bolstered-with-new-funding/dMI7OQCL5T8e5vTBfrOKJM/

[10] Julia Arenstam, "Louisiana's Houma-Terrebonne Airport Gets $1.3M Drone Grant," The Houma Courier, June 20, 2019.

https://www.govtech.com/products/Louisianas-Houma-Terrebonne-Airport-Gets-1-3M-Drone-Grant.html

[11] "Dayton tech firm wins funding for 'Air Uber' system," July 16, 2018.
https://www.bizjournals.com/dayton/news/2018/07/16/dayton-tech-firm-wins-funding-for-air-uber-system.html

[12] The White House, "FACT SHEET: New Commitments to Accelerate the Safe Integration of Unmanned Aircraft Systems," August 2, 2016.
https://obamawhitehouse.archives.gov/the-press-office/2016/08/02/fact-sheet-new-commitments-accelerate-safe-integration-unmanned-aircraft

[13] Rick Smith, "NCSU, partners land $24M grant to develop first platform linking drones, 5G wireless technology," September 18, 2019.
https://www.wraltechwire.com/2019/09/18/ncsu-partners-land-24m-grant-to-develop-first-platform-linking-drones-5g-wireless-technology/

[14] "Drone research possibilities soar thanks to NSF grant," October, 2018.
https://blog.smu.edu/smumagazine/2018/10/drone-research-possibilities-soar-thanks-to-nsf-grant/

[15] Daniel DeVries, "Major NSF Grant to Fund Drone Research in Antarctica," July 17, 2019.
https://www.colgate.edu/news/stories/major-nsf-grant-fund-drone-research-antarctica

[16] "NSF awards $6 million to OSU-led collaboration to develop weather research UAVs," August 7, 2015.
https://news.okstate.edu/articles/communications/2015/nsf-awards-6-million-osu-led-collaboration-develop-weather-research-uavs.html

[17] "OSU drone project wins $1.49 million NSF grant," December 6, 2019.

https://news.okstate.edu/articles/engineering-architecture-technology/2019/osu_researchers_awarded_nsf_nri_grant_to_study_wind_using_drones_and_enhance_drone_operation_safety.html

[18] "National Science Foundation grant to fund drone advances at UNT," August 14, 2017.

https://news.unt.edu/news-releases/national-science-foundation-grant-fund-drone-advances-unt

[19] Michael Miller, "Ohio panel recognizes UC's drone work," July 18, 2018.

https://magazine.uc.edu/editors_picks/recent_features/uavgrant.html

[20] "Researcher Leads Project To Explore Use Of Drones To Fight Wildfires," January 23, 2019.

https://news.gsu.edu/2019/01/23/georgia-state-grant-to-study-using-drones-to-fight-wildfires/

[21] "NSF grant supports development of robotic frost protection in orchards," March 25, 2019.

https://news.psu.edu/story/565198/2019/03/25/research/nsf-grant-supports-development-robotic-frost-protection-orchards

[22] "Researchers receive $1.25 million grant to map marine ecosystems with drones," September 7, 2018.

https://www.eurekalert.org/pub_releases/2018-09/uocf-rr090718.php

[23] Siddharth Vodnala, "U.S. Army grant supports development of hydrogen-powered Unmanned Aerial Vehicle," July 18, 2019. https://news.wsu.edu/2019/07/18/us-army-grant-supports-development-hydrogen-powered-unmanned-aerial-vehicle/

[24] Award Abstract #1814748, "NeTS: Small: Free Space Optics as Backhaul and Energizer for Drone-assisted Networking," National Science Foundation, August 22, 2018. https://www.nsf.gov/awardsearch/showAward?AWD_ID=1814748

[25] Michaela Jarvis, "New Drone Study to Help Farmers Brace for Climate Change," August 28, 2019. https://news.erau.edu/headlines/new-drone-study-to-help-farmers-brace-for-climate-change

[26] Award Abstract #1837779, "CPS: Small: Novel Algorithmic Techniques for Drone Flight Planning on a Large Scale," National Science Foundation, September 12, 2018. https://www.nsf.gov/awardsearch/showAward?AWD_ID=1837779&HistoricalAwards=false

[27] Grant No.2017-67007-26148, "Development of pipeline & tools for drone-based canopy phenotyping in crop breeding," National Institute of Food and Agriculture, Feb 1, 2017. https://portal.nifa.usda.gov/web/crisprojectpages/1011761-development-of-pipeline-and-tools-for-drone-based-canopy-phenotyping-in-crop-breeding.html

[28] Award Abstract #1931962, "CPS:Medium:Interactive Human-Drone Partnerships in Emergency Response Scenarios," National Science Foundation, September 12, 2019. https://www.nsf.gov/awardsearch/showAward?AWD_ID=1931962

[29] "NSF grant to help NAU grow and share drone technology," May 11, 2016.

http://news.nau.edu/nsf-grant-help-nau-grow-share-drone-technology/

[30] Project Grant FAIN 80NSSC18K1519, NASA, Jan 01, 2019.

https://www.usaspending.gov/#/award/ASST_NON_80NSSC18K1519_8000

[31] Christopher Allen (PI), Mark Ewing, and Shariar Keshmiri, "Final Research Report: Multichannel Sense-and-Avoid Radar for Small UAVs," NASA Award Number: NNX13AB76A, University of Kansas, 2013.

https://nari.arc.nasa.gov/sites/default/files/Allen_LEARN%20Final%20Report%20-%20Kansas%20-%20Jan%202015%20%20%28v2%29.pdf

[32] Jason Reaganon, "Multi-million Dollar Grant Will Fund Virginia Unmanned Systems Project," October 18, 2019.

https://dronelife.com/2019/10/18/multi-million-dollar-grant-will-fund-virginia-unmanned-systems-project/

[33] Chris Adam, "Multi-sensor drone technology for plant phenotyping receives $4.5 million from energy agency," December 2, 2019.

https://www.purdue.edu/newsroom/releases/2019/Q4/multi-sensor-drone-technology-for-plant-phenotyping-receives-4.5-million-from-energy-agency.html

[34] Tom Henderson, "Drone-inspection company SkySpecs raises $17 million funding round," December 21, 2019.

https://www.crainsdetroit.com/technology/drone-inspection-company-skyspecs-raises-17-million-funding-round

[35] Alan Boyle, "Microsoft TechSpark grants $100K to North Dakota's Airtonomy drone startup," April 23, 2019.
https://www.geekwire.com/2019/microsoft-techspark-grants-100k-north-dakotas-airtonomy-drone-startup/

[36] Lucas Matney, "Impossible Aerospace raises $9.4M to sell drones stuffed with battery cells," September 10, 2018.
https://techcrunch.com/2018/09/10/impossible-aerospace-raises-9-4m-to-sell-drones-stuffed-with-battery-cells/

[37] Josh Constine, "Anti-drone radio wave startup SkySafe secures $11.5M from Andreessen," July 20, 2017.
https://techcrunch.com/2017/07/20/skysafe/

[38] Jonathan Shieber, "WhiteFox Defense lands $12 million as the demand for drone defense technologies intensifies," December 5, 2018.
https://techcrunch.com/2018/12/05/whitefox-defense-lands-12-million-as-the-demand-for-drone-defense-technologies-intensifies/

[39] Gettinger, Dan. "Drones in the Defense Budget: Navigating the Fiscal Year 2018 Budget Request," Center for the Study of the Drone at Bard College, October 24, 2017.
https://dronecenter.bard.edu/projects/defense-spending-on-drones/drones-2018-defense-budget/

[40] Gettinger, Dan. "Study: Drones in the FY 2019 Defense Budget," Center for the Study of the Drone at Bard College, 2018.

https://dronecenter.bard.edu/drones-in-the-fy19-defense-budget/

[41] Elizabeth Gibney, "Europe's controversial plans to expand defence research," May 22, 2019.

https://www.nature.com/articles/d41586-019-01567-y

[42] Zhang Xia, "Chinese Drone Firm MicroMultiCopter Secures USD30 Million in Pre-IPO Funding," Nov 28 2017.

https://www.yicaiglobal.com/news/chinese-drone-firm-micromulticopter-secures-usd30-million-in-pre-ipo-funding

[43] Catherine Shu, "Chinese Drone Startup Yuneec Grabs $60M In Funding From Intel," August 27, 2015.

https://techcrunch.com/2015/08/27/yuneec/

Product website: https://www.yuneec.com/en_US/home/intel-realsense-technology-3429.html

[44] Li Tao and Yingzhi Yang, "Your Chinese takeaway is ready to be delivered by drone," May 30, 2018.

https://www.scmp.com/tech/china-tech/article/2148350/your-chinese-takeaway-ready-be-delivered-drone

[45] Philip Butterworth-Hayes, "EU-China APP Drone Workshop highlights UTM progress in China," UAS Traffic Management News, June 12, 2018.

https://www.unmannedairspace.info/uncategorized/eu-china-app-drone-workshop-highlights-utm-progress-china/

[46] Marco Margaritoff, "China's New Drone Innovation Center in Shanghai Will Boost Developing Technologies," August 31, 2018. https://www.thedrive.com/tech/23266/chinas-new-drone-innovation-center-in-shanghai-will-boost-developing-technologies

[47] "International Opportunities for Danish Drone Researchers, Innovation Centre Denmark: Mapping of Drone Research Environments in Seven Countries," December, 2017. https://ufm.dk/en/publications/2017/international-opportunities-for-danish-drone-researchers

[48] China Power Team, "Is China at the forefront of drone technology?," May 29, 2018. https://chinapower.csis.org/china-drones-unmanned-technology/

[49] "AI radar system that can spot miniature drones 3 kilometers away," July 18, 2019. https://www.sciencedaily.com/releases/2019/07/190718085310.htm

[50] Evan Ackerman, "South Korea Prepares for Drone vs. Drone Combat," April 1, 2015. https://spectrum.ieee.org/automaton/robotics/drones/south-korea-drone-vs-drone

[51] "South Korea tests drone delivery in remote regions," July 31, 2019. https://techxplore.com/news/2019-07-south-korea-drone-delivery-remote.html

[52] Alex Douglas, "South Korea's first government drone project completed with SimActive," September 19, 2018.
https://www.commercialdroneprofessional.com/south-koreas-first-government-drone-project-completed-with-simactive/

[53] Kim Min-Seok, "The South Korean military is catching up with a new drone army," February, 2018.
https://www.the-security-times.com/the-south-korean-military-is-catching-up-with-a-new-drone-army/

[54] Captain Sukjoon Yoon, Republic of Korea Navy (Ret.), "A Perspective on South Korea," May, 2016.
http://drones.cnas.org/reports/a-perspective-on-south-korea/

[55] Hyonhee Shin, "South Korea to develop anti-drone system amid standoff with North," September 16, 2019.
https://www.reuters.com/article/us-northkorea-southkorea-drone/south-korea-to-develop-anti-drone-system-amid-standoff-with-north-idUSKBN1W20G5

[56] Jung Suk-yee, "Fear of Drone-based Terrorist Attacks Spreading in South Korea," September 18, 2019.
http://www.businesskorea.co.kr/news/articleView.html?idxno=36087

[57] Rob Gilhooly, "Entering the drone age: Japan seeks to tap into the potential of unmanned flying vehicles," January 20, 2018.

https://www.japantimes.co.jp/life/2018/01/20/digital/entering-drone-age-japan-seeks-tap-potential-unmanned-flying-vehicles/

[58] Jeremiah Karpowicz, "Drone Fund Showcases Numerous Innovative UAS Solutions from Japan," October 13, 2019.

https://www.commercialuavnews.com/forestry/drone-fund-uas-solutions-japan

[59] Emma Cosgrove, "Japan's Nileworks Raises $7.1m for Crop Spraying Drones," October 17, 2017.

https://agfundernews.com/japans-nileworks-raises-7m-automated-sprayer-drones.html

[60] Sven Eriksson and Maths Lundin, "The Drone Market in Japan," EU-Japan Centre for Industrial Cooperation, September, 2016.

https://www.eubusinessinjapan.eu/sites/default/files/drone_market_in_japan.pdf

[61] Lieutenant General Masayuki Hironaka, Japan Air Self-Defense Force (Ret.), "A Perspective on Japan," May, 2016.

http://drones.cnas.org/reports/a-perspective-on-japan/

Made in United States
Cleveland, OH
04 November 2026

47617536R10037